Kenny's Favorite Things to Know

Kenny's Favorite Things to Know

By Kenny 'K3' Rochon, III
and Ken 'Dr. Smiley' Rochon, Jr., PhD (Daddy)

ISBN: 978-1-64810-054-3

PRINTED IN THE UNITED STATES OF AMERICA

This is dedicated my Mom and Dad and Grandpa and Grandma Ruthann Aunt Lizy, Uncle Alex, Caroline, Stephanie, Uncle Stevie, Aunt Sharron, Matt Rochon, Andrew Rochon, Aunt Jen, Uncle Patrick, Kay Kay Herwig, Brendan Herwig.

To my friends: Ally, Ayden, Drake, Nico, Jackson, Jayden, and Nico.

To my teachers Mrs. Bray, Mrs. Braun, Mr Croushler & my principal Mr. Martin

-Kenny

This is dedicated to my favorite people in the world. My son who lights up my life and inspires me to publish a book a year for him and my loving wife who is the best mother in the world to our son. Raising a child with love is the way to ultimately bring more smiles to the world.

Also, a special thanks to Carolyn Sheltraw for her help with layout, Tara Hannon for the front cover illustration. Al Granger for his hand in making another book (almost 200) a reality and a super fun book to help fund Kenny's education.

- Ken 'Dr. Smiley' Rochon, Jr., PhD

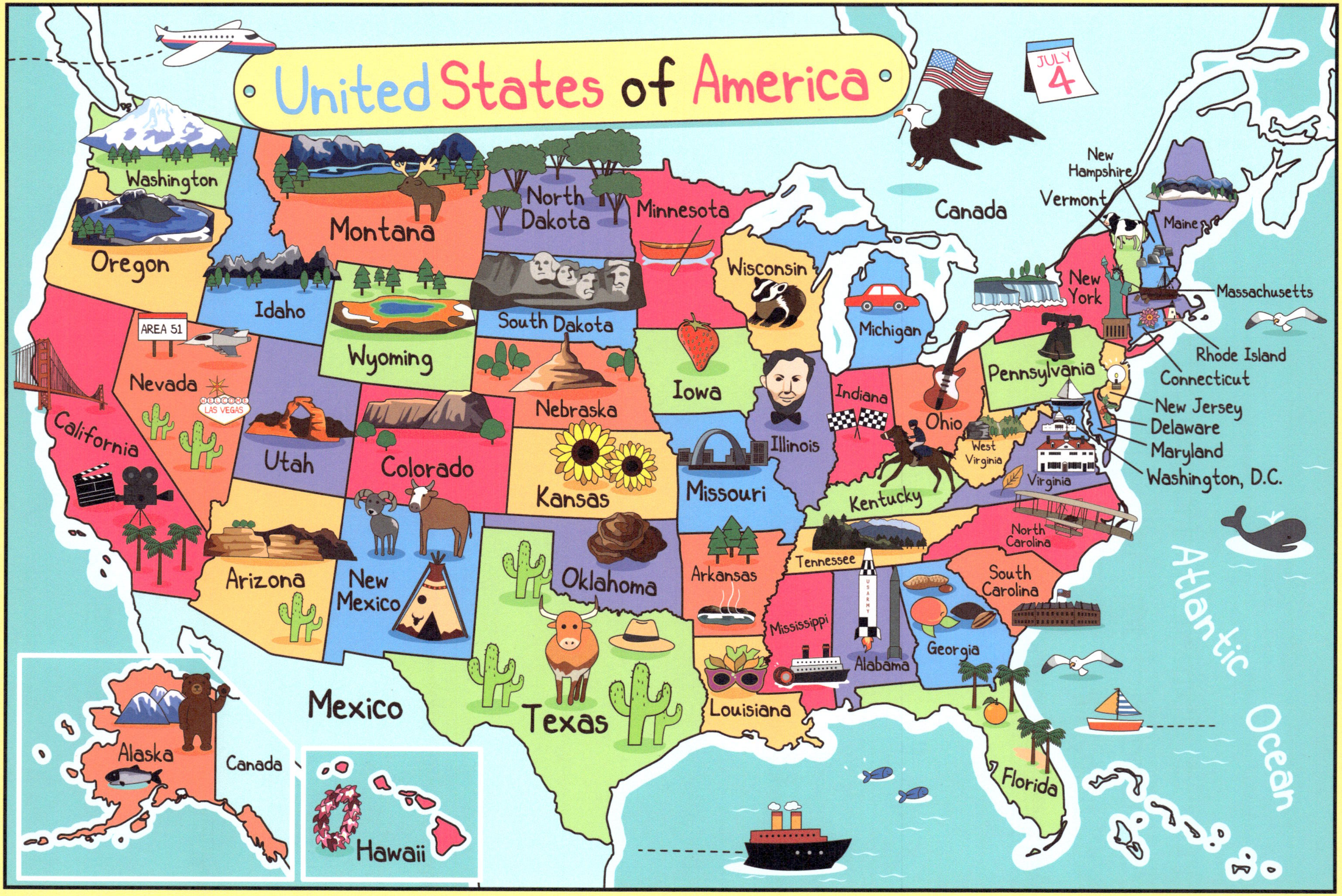
United States of America
JULY
4
Canada
Washington
Oregon
California
Nevada
AREA 51
LAS VEGAS
Idaho
Montana
Wyoming
Utah
Arizona
Colorado
New Mexico
North Dakota
South Dakota
Nebraska
Kansas
Oklahoma
Texas
Minnesota
Iowa
Missouri
Arkansas
Louisiana
Wisconsin
Illinois
Michigan
Indiana
Ohio
Kentucky
Tennessee
Mississippi
Alabama
Georgia
Florida
South Carolina
North Carolina
Virginia
West Virginia
Pennsylvania
New York
Vermont
New Hampshire
Maine
Massachusetts
Rhode Island
Connecticut
New Jersey
Delaware
Maryland
Washington, D.C.
Atlantic Ocean
Mexico
Alaska
Canada
Hawaii

AMERICA

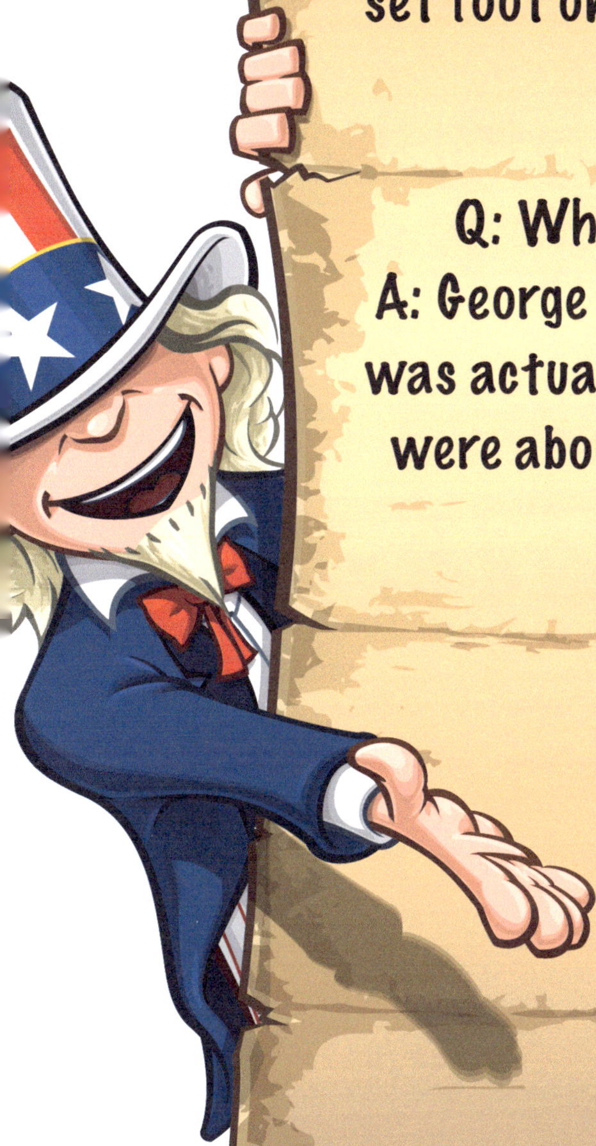

Q: Who discovered America?
A: Leif Erikson FYI: Christopher Columbus never set foot on what we now call the United States in all four of his voyages

Q: Who was the first American president?
A: George Washington (elected). Peyton Randolph was actually the first and 3rd President and there were about a dozen prior to George Washington

Q: When was the declaration of Independence signed?
A: August 2nd, 1776

Q: What day of the year is Independence Day?
A: July 4th

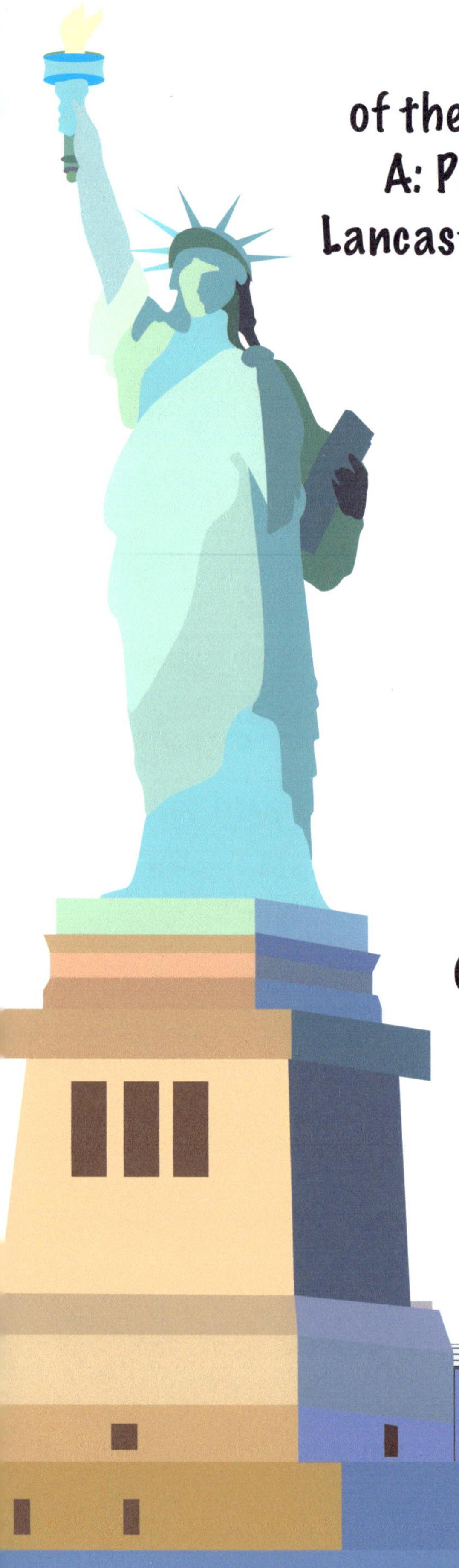

Q: What city was the first capital of the United States? Bonus: What other cities?
A: Philadelphia. Bonus: Annapolis, Baltimore, Lancaster, New York City, Princeton, Trenton, York

Q: What is the Capital of the United States of America?
A: Washington D.C.

Q: What was the first state?
A: Delaware

Q: What is the largest state in America?
A: Alaska

Q: What is the smallest state in America?
A: Rhode Island

Q: The Statue of Liberty came from which country?
A: France

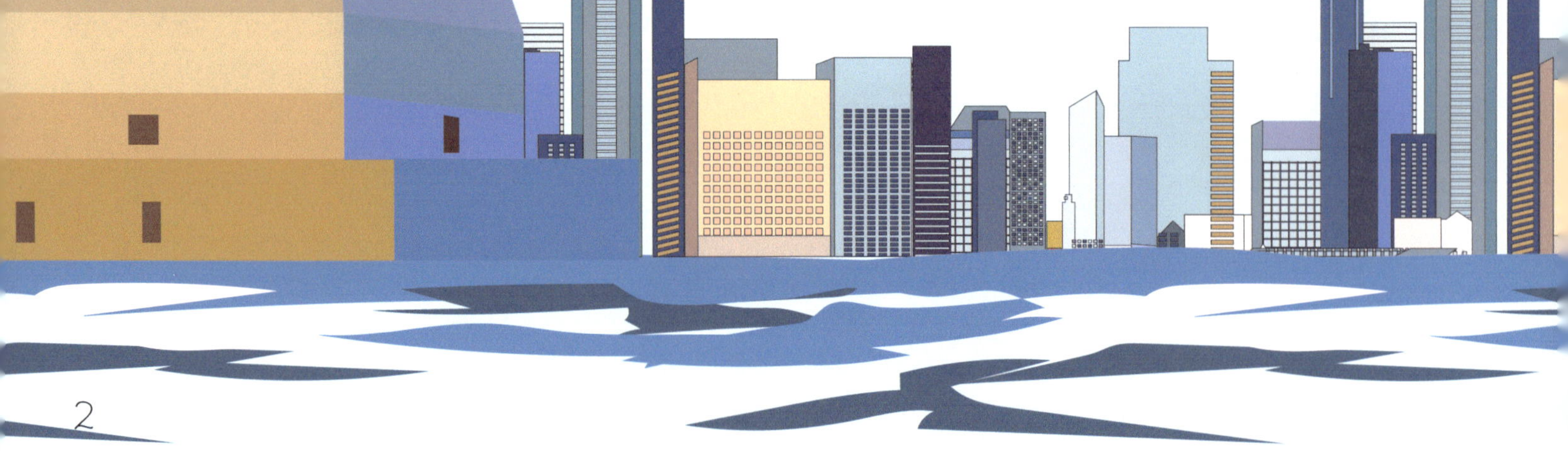

Q: Where does the President of the United States live while in office?
A: The White House

Q: American pioneer, John Chapman, was famous for planting which kind of tree?
A: Apple Trees

Q: Which state is famous for Hollywood?
A: California

Q: The world's largest man-made oyster reef was created in what state?
A: Maryland

Q: What is the name of the oldest living tree?
Bonus: How old and where are they?
A methuselah. Bonus: 4,843 years old, California

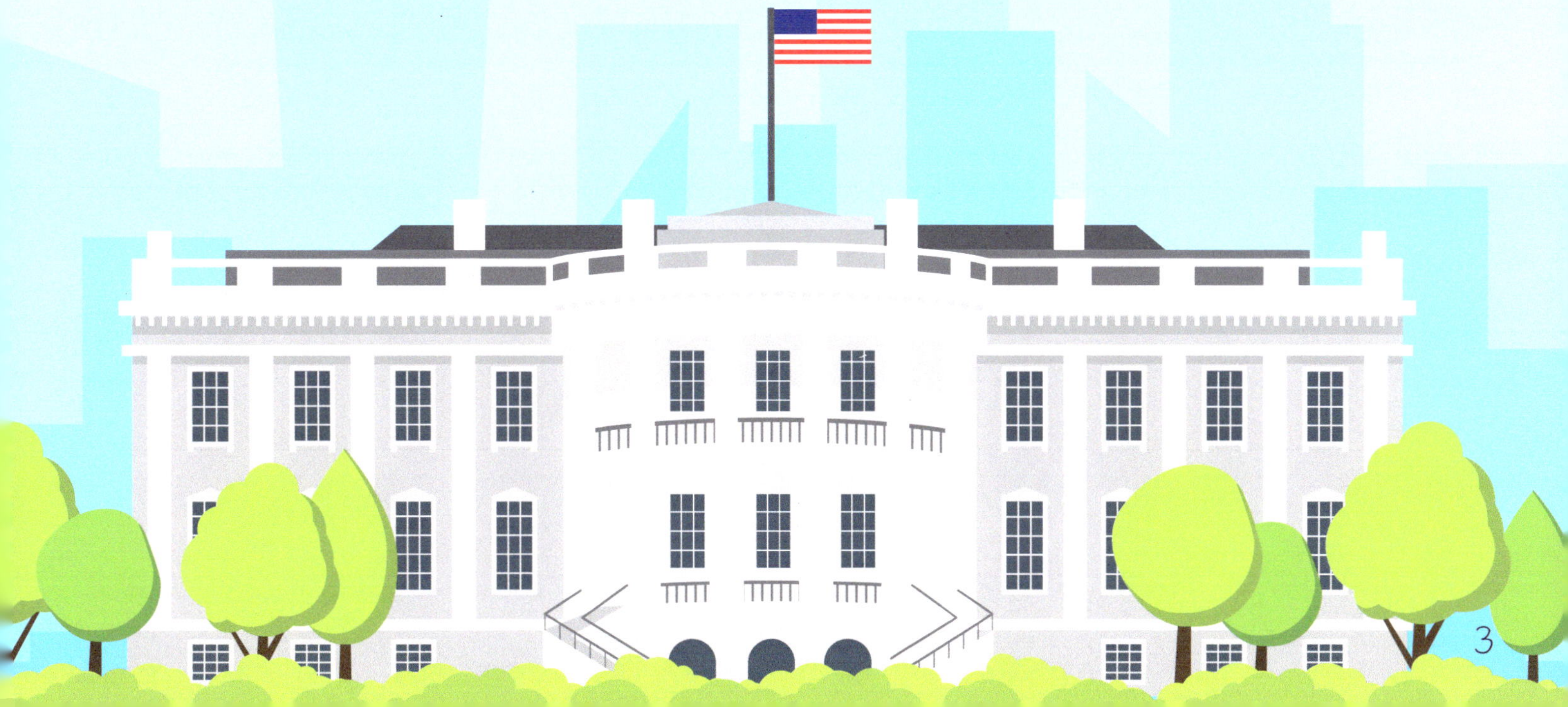

Q: What are the only two U.S. states that can grow coffee commercially?
A: Hawaii and California

Q: How many Great Lakes are there?
A: Five

Q: What letter is not included in any of the 50 U.S. state names?
A: The letter Q

Q: Who was the 16th president of the United States?
A: Abraham Lincoln

Q: Which U.S. state has the longest cave system in the world?
A: Kentucky

Q: What is the only state that borders just one other state?
A: Maine

Q: Which two U.S. states don't observe Daylight Savings Time?
A: Arizona and Hawaii

Q: Which state was the 50th state to join the union?
A: Hawaii

Q: Samuel Tilden, Grover Cleveland, Al Gore, and Hillary Clinton share what distinction among U.S. president candidates?
A: They all won the popular vote, but lost the electoral-college vote

Q: Who was the only U.S. President to resign?
A: Richard Nixon

Q: What three word phrase is written on Martin Luther King's gravestone?
A: Free at Last

Q: How many stars and stripes on the flag?
Bonus: Why? Bonus 2: How many red stripes?
A: 50 stars, 13 stripes.
Bonus 50 States, 13 original colonies. 7 red stripes

ANIMALS

Q: How many bones do sharks have?
A: Zero

Q: How many pairs of wings do bees have?
A: Two

Q: How many noses does a slug have?
A: Four

Q: How many legs does a spider have?
A: Eight

Q: How many legs does a lobster have?
A: 10 (8 walking legs and 2 larger claw legs)

Q: What percentage of Animals are extinct?
A: 99%

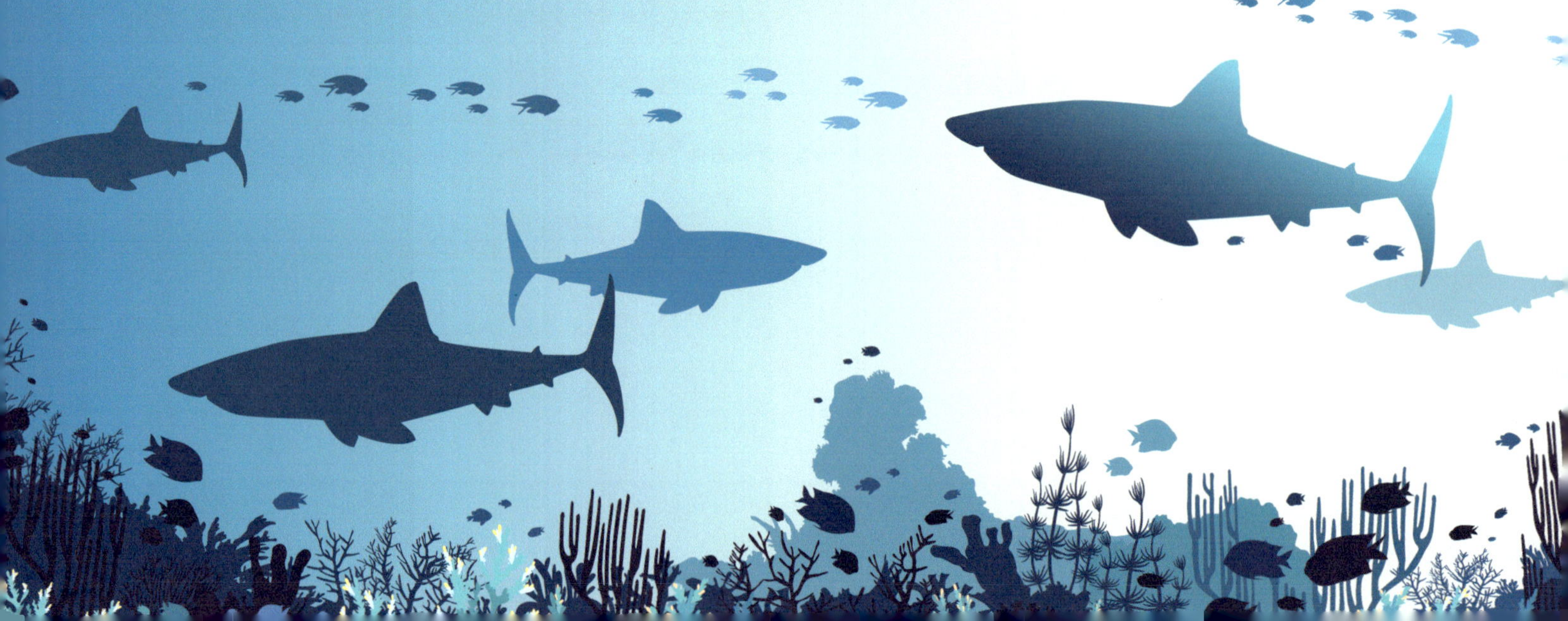

Q: What percentage of Animals are invertebrates?
A: 95%

Q: What is the deadliest animal in the world?
A: Hippopotamus.

Q: Which animal kills the most after humans?
Bonus: How many? Bonus 2: What diseases?
A: Mosquito. Bonus: they afflict an estimated 700 million and kill roughly 725,000 a year. Bonus 2: Malaria, Chikungunya, Dengue Fever, Encephalitis, Elephantiasis, and Yellow Fever.

Q: What is the fastest land animal? Bonus: Speed in mpg?
A: The Cheetah. They have set record speeds near 70 mph.

Q: What is the fastest aquatic animal?
Bonus: Speed in mph? Bonus 2: Shark & Dolphin mph?
A: The sailfish. Bonus: It can reach speeds of up to 68 mph.
Bonus 2: Shark 44 mph, Dolphin 37 mph.

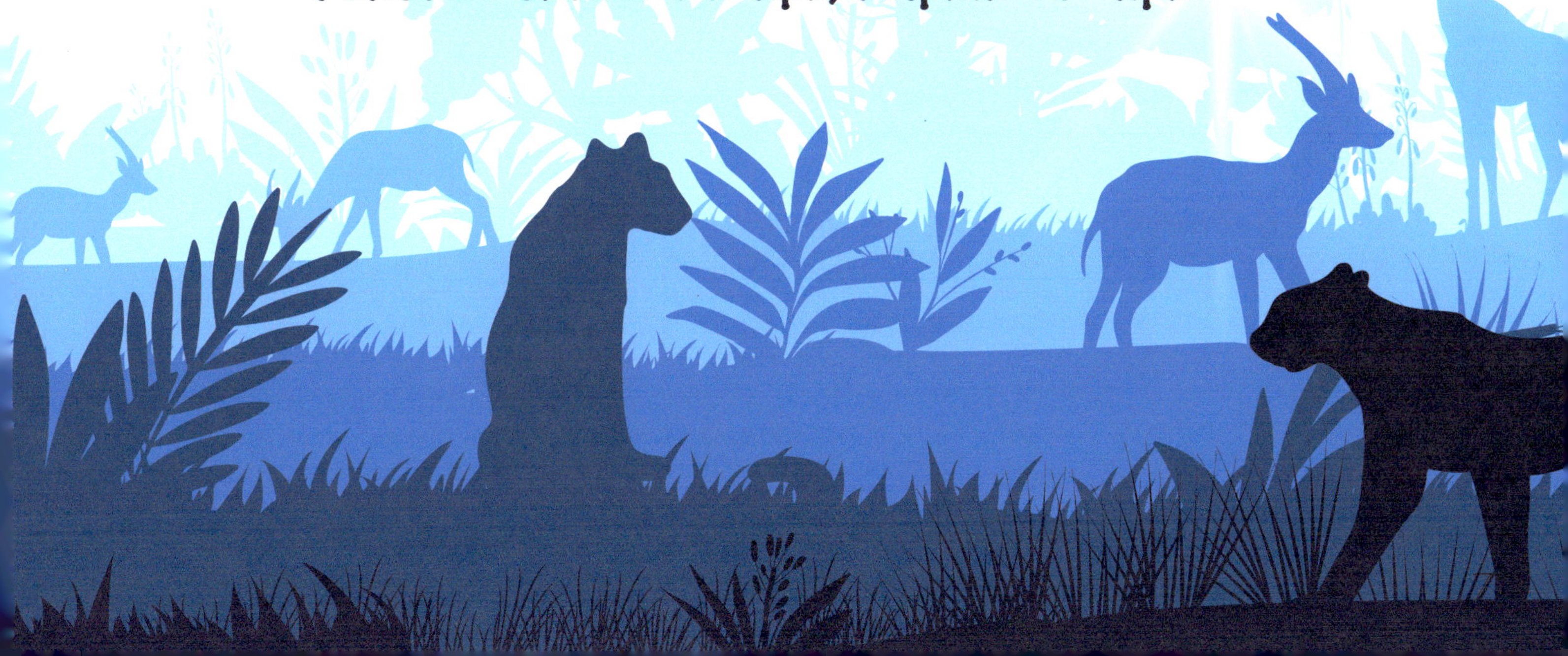

Q: What is the largest land animal on Earth?
A: Elephant

Q: What is the largest animal on Earth?
Bonus: Weight?
A: Blue Whale
Bonus: Length - 90 ft. and Weighs about 400,000 pounds.

Q: What is the strongest living creature on Earth?
A: Gonorrhea Bacterium is pound for pound the strongest living creature on Earth. Capable of pulling up to 100,000 times its own body weight.

Q: Which animal is the tallest in the world?
Bonus: How tall?
A: Giraffe. Bonus: 15 - 20 feet.

Q: What do you call a group of giraffes?
A: A Tower

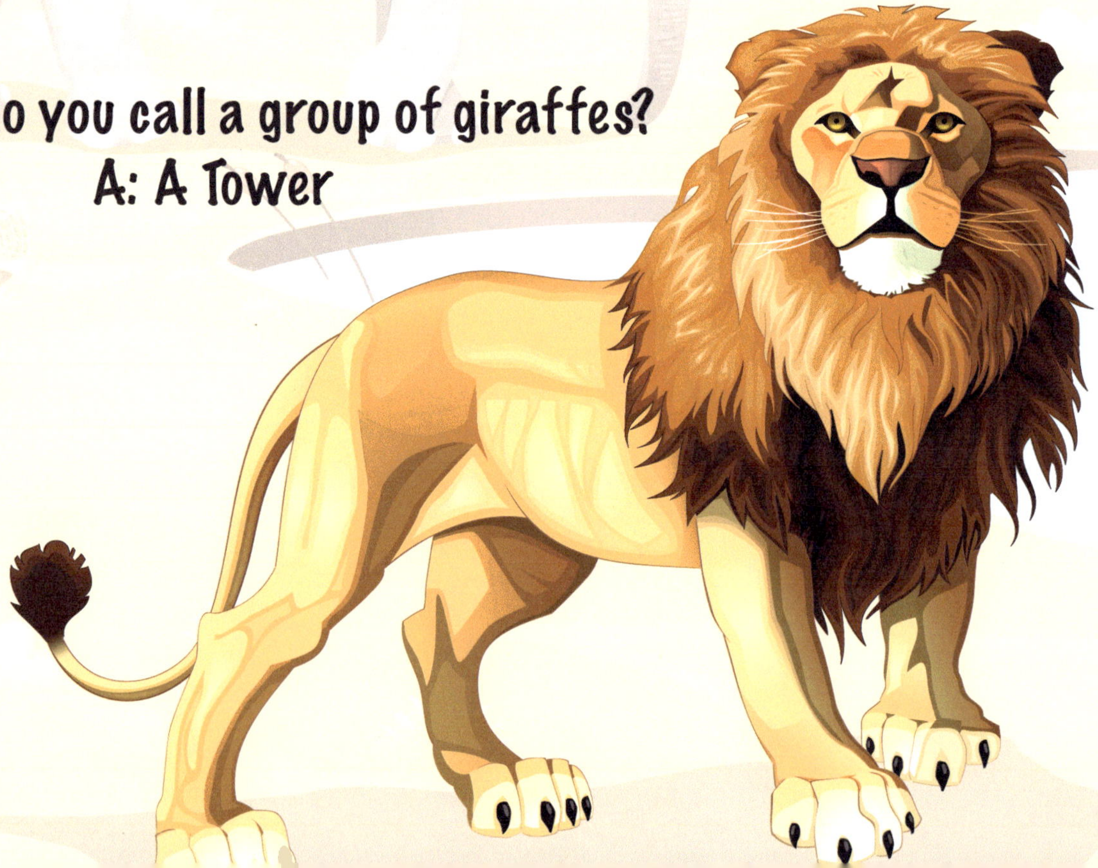

Q: Which animal has no vocal cords?
A: Giraffe

Q: What is a group of lions called?
A: A Pride

Q: What is a group of tigers called?
A: An Ambush or Streak

Q: Are worker bees male or female?
A: Female

Q: What is a doe?
A: A female deer

Q: Name a mammal that can't jump.
A: Elephant, sloth, hippo, rhino

Q: How many hearts does an Octopus have?
A: 3

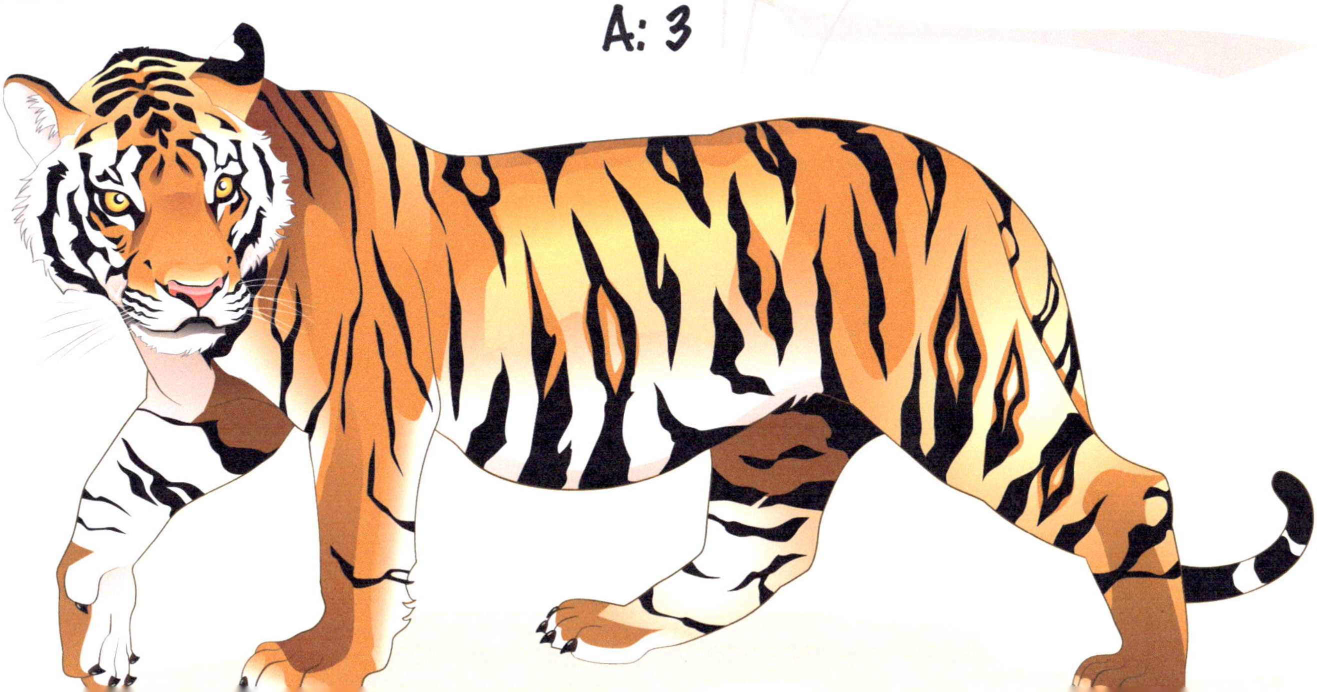

Q: What food do pandas eat?
A: Bamboo

Q: If you suffer from arachnophobia, which animal are you scared of?
A: Spiders

Q: What animal is considered the best jumper in the world?
Bonus: How high?
A: Flea. Bonus 220 times their own body length.

Q: What is the loudest animal on earth?
A: The sperm whale

Q: Which mammal lives the longest?
A: Bowhead whale. They can live up to 200 years!

Q: What animal has the strongest bite in the world? Bonus: What is the psi (pressure per inch)
A: Nile Crocodile. Bonus: 5,000psi. Shark 1,350psi, Jaguar 1,500psi, Hippo 1,800psi, Humans are 200psi.

Q: How can you get African Sleeping Sickness?
Bonus: What happens?
A: Tsetse Fly is a bloodsucking bug.
Bonus: It transmits protozoan parasites that have a neurological and meningoencephalitic symptoms including behavior changes and poor coordination.

ARTS

Q: What is the largest museum in the world?
Bonus: Where?
A: The Louvre. Bonus: Paris, France.

Q: What is the largest museum, education
and research complex? Bonus: How many?
A: The Smithsonian Institution.
Bonus: 19 museums and galleries.

Q; What is the most expensive painting sold in auction?
Bonus: How much?
A: Salvator Mundi. Bonus: $450.3 million.

Q: Who painted the Mona Lisa?
Bonus: Where is it displayed?
A: Leonardo DaVinci.
Bonus: The Louvre, Paris, France.

Q: Who created the Statue of David?
Bonus: Where is it located?
A: Michelangelo. Bonus: Florence, Italy. Considered the most
famous and known sculpture in the world.

COLORS

Q; What is the most popular color in the world?
Bonus: What percent picked this color?
A: Blue. Bonus: 40% (purple was second at 14%)

Q: What color is a polar bear's skin?
A: Black

Q: What colors in a rainbow order from top to bottom?
A: ROYGBIV: Red, orange, yellow, green, blue, indigo, violet.

Q: What color are the stars in the American flag?
A: White

Q: What is the most popular color of car?
Bonus! What is the 2nd most popular color?
A: White. Bonus: Silver

Q: What color attracts mosquitoes the most?
A: Dark Blue. Instead wear light colored, loose fitting long sleeves.

Q: What is it called if you fear colors?
A: Chromophobia.

Q: What eliminates Black and White
from being considered a color?
A: They do not have specific wavelengths.
Color is a visible light with a specific wavelength.

CULTURAL

Q: Which continent is the most linguistically diverse in the world? Bonus: How many languages?
A: Africa. Bonus: 800 to 1500 of the world's languages

EARTH

Q: How old is Earth?
A: 4.5 Billion years

Q: Circumference of Earth?
A: 24,901 miles

Q: What are the three layers of Earth?
A: Core, Mantle, and Crust

Q: How fast is Earth spinning through space?
A: Over 1,000 miles per hour

Q: Hottest spot on Earth? Bonus: Other Ho
A: Death Valley, California set the world record of 134° F was recorded July 10, 2013. Bonus: Lut Desert, Iran measured 159.3° F 2005 by NASA's Aqua Satellite. Aziziyah, Libya, 136° F degrees Fahrenheit. 2005

Q: What is the largest living structure on Earth?
A: The Great Barrier Reef world's largest coral reef composed of over 2900 individual reefs and 900 miles stretching 1,400 miles (and 133,000 sq. mi.). Located in the Coral Sea off the coast of Queensland, Australia.

Q: Where are the world's most active volcano located?
A: Hawaii

Q: What is the name of molten rock after a volcanic eruption?
A: Lava

Q: What is the hardest natural substance?
A: A diamond

Q: How far is the Moon from Earth in miles?
A: 238,900 mi

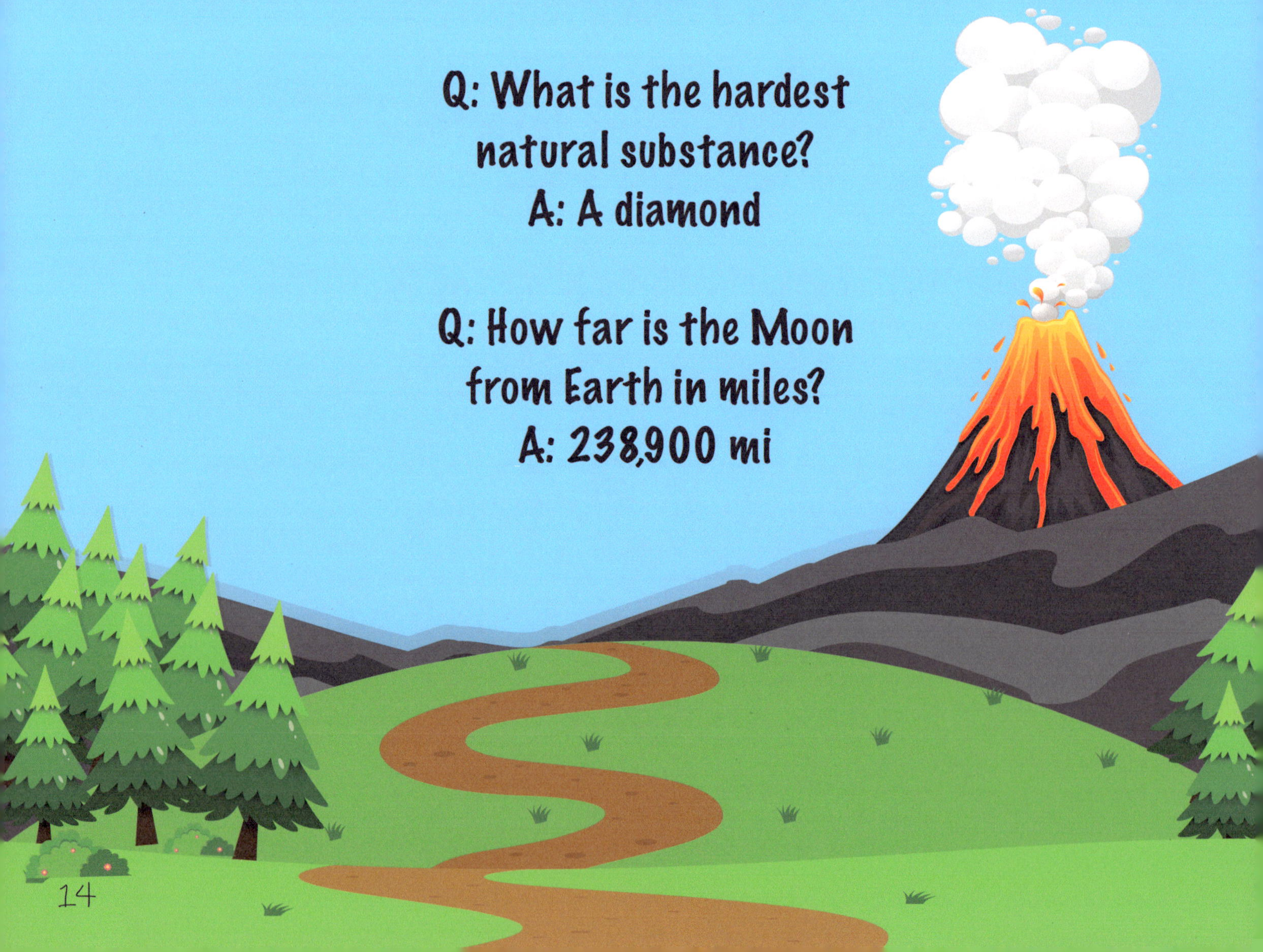

ENTERTAINMENT & MOVIES

Q: On which hand did Captain Hook have his hook?
A: Left

Q: What's the name of the toy cowboy in Toy Story?
A: Woody

Q: What type of animal is Baloo in The Jungle Book?
A: A bear

Q: What type of fish is Nemo?
A: A clownfish

Q: Who is Mickey Mouse's girlfriend?
A: Minnie Mouse

Q: What was the first Disney animated film based on the life of a real person?
A: Pocahonntas

Q: What is the name of the fairy in Peter Pan?
A: Tinkerbell

Q: Who lives in a trashcan on Sesame Street?
A: Oscar the Grouch

Q: What movie are characters looking for a brain, a heart, and courage?
A: Wizard of Oz

FOOD

Q: Healthiest food in the world? BONUS: Name 4 more.
A: Spinach. Bonus: Black Beans, Walnuts, Beets, & Alvocado

Q: What fruit floats because 25% of its volume is air?
A: Apple

Q: What food serves as the base for guacamole?
A: Avocado

Q: What food is the most ordered in America?
A: Fried chicken

Q: What country produces the most coffee in the world?
A: Brasil

Q: What is the world record for number of hotdogs eaten in one sitting?
A: 74

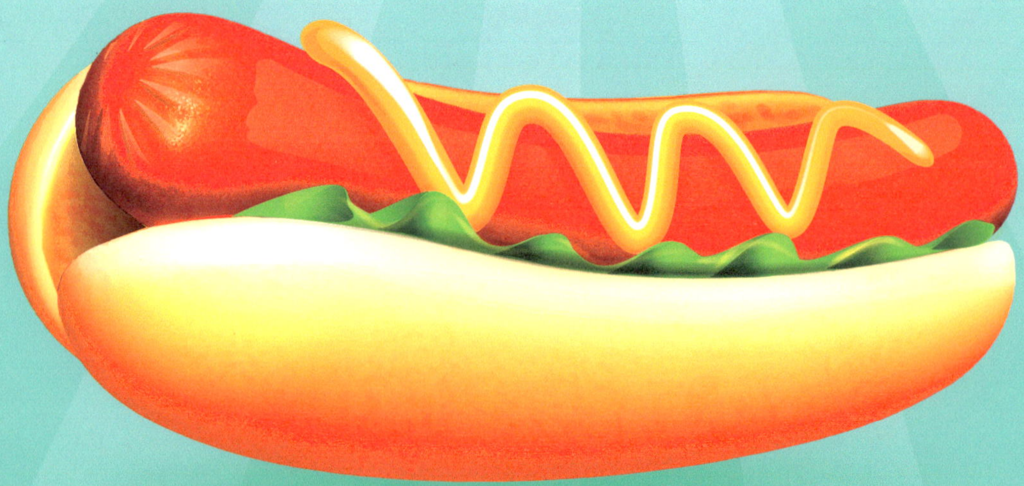

Q: What is the only fruit that has its seeds on the outside?
Bonus: How many?
A: Strawberries. Bonus: 200

Q: What contains more sugar, strawberries or lemons?
A: Lemons

Q: Can you name the largest chocolate
manufacturer in the United States?
A: Hershey's. Bonus! In which state can you find
the city of Hershey? Pennsylvania.

Q: Which country is the largest producer of vanilla?
A: Madagascar

Q: What do you eat as a vegetarian?
A: Plant-based foods, like vegetables,
fruits, beans, soy, grains, rice and nuts.

GEOGRAPHY

Q: How many continents are there? What are the names?
A: Africa, Antarctica, Asia, Australia, Europe, North America & South America

Q: What is the largest continent?
A: Asia

Q: What is the driest continent?
A: Antartica

Q: What is the largest country in the world?
A: Russia

Q: What is the smallest country in the world?
A: Vatican City

Q: Where is the Great Pyramid of Giza?
A: Egypt

Q: Which country is home to the kangaroo?
A: Australia

Q: What is the largest ocean on Earth?
A: The Pacific

Q: What is the smallest ocean in the world?
A: The Artic

Q: What is the tallest mountain in the world?
A: Mount Everest

Q: In which capital city of Europe would you find the Eiffel Tower? Bonus: It was originally intended for what city?
A: Paris. Bonus: Barcelona

Q: Which two countries share the longest border?
A: The United States and Canada

Q: What is the longest river in the world?
A: The Nile

Q: "Stars and Stripes" is the nickname of the flag of which country?
A: United States of America

Q: What city has the largest population in the world? Bonus: population. Bonus 2: 2nd, 3rd?
A: Tokyo. Bonus: 37 million. Delhi – 30 million, Shanghai – 26 million

Q: What country has the most vending machines per capita?
A: Japan

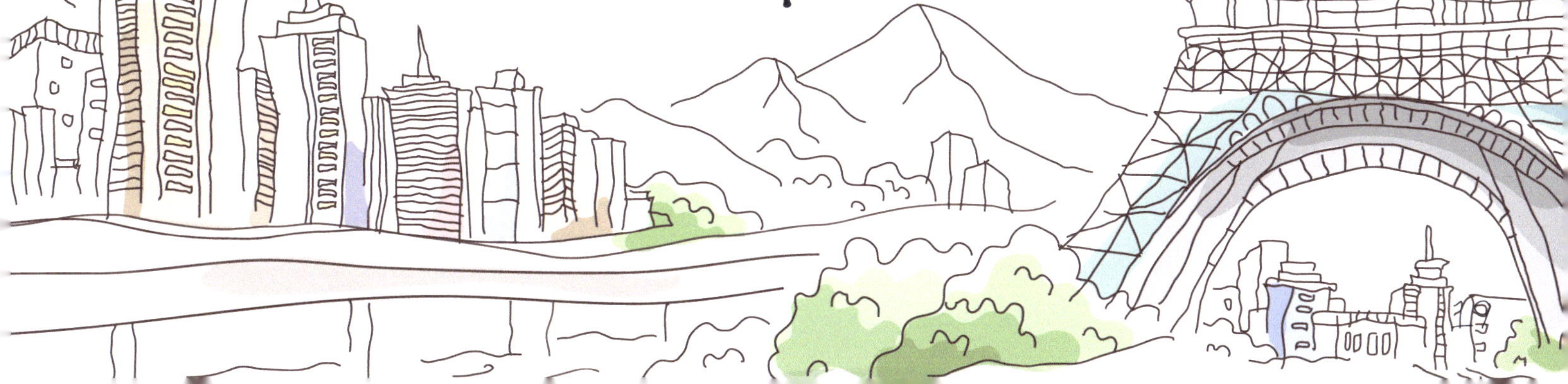

HUMAN BODY

Q: What is the essential gas we need to breath?
Bonus: Chemical Symbol
A: Oxygen. Bonus O

Q: Water is what percent of the human body?
A: 60%

Q: How many bones in the human body?
A: 206

Q: What bone are babies born without?
A: Knee cap

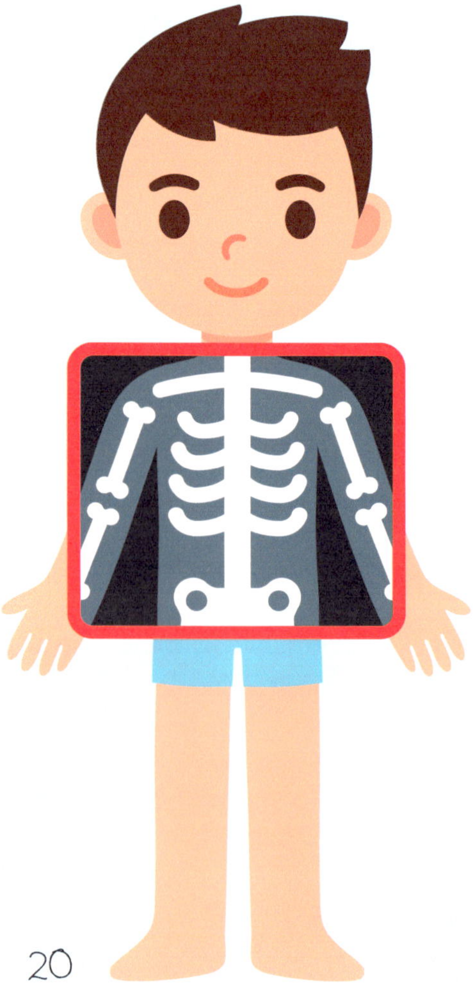

Q: How many times does your heart beat a day?
A: About 100,000 times

Q: How many cells are in your body?
A: Over 10 trillion cells

Q: How fast is a nerve impulse from the brain?
A: 274 mph

Q: What part of the brain deals with hearing and language?
A: Temporal lobe

Q: What is the longest and strongest bone?
A: The Femur

Q: What is the smallest bone in the human body?
A: Middle ear

Q: How long is the human intestine?
A: 17 feet

Q: How many miles of blood vessels are there in the human body?
A: 59,650 miles

Q: How long does a human red blood cell survive?
A: 120 days

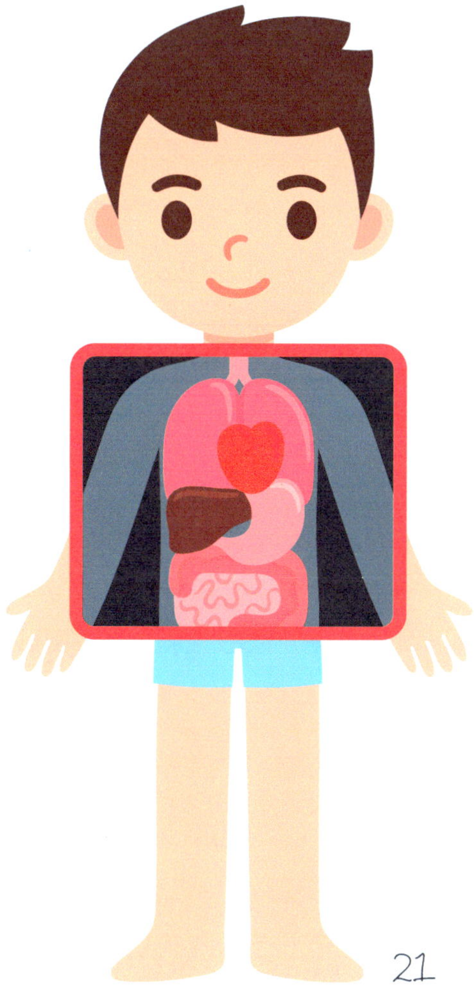

Q: How many teeth does the adult human have?
A: 32. Bonus Children have 20

Q: What vitamin prevents scurvy suffered by pirates on long voyages?
A: Vitamin C

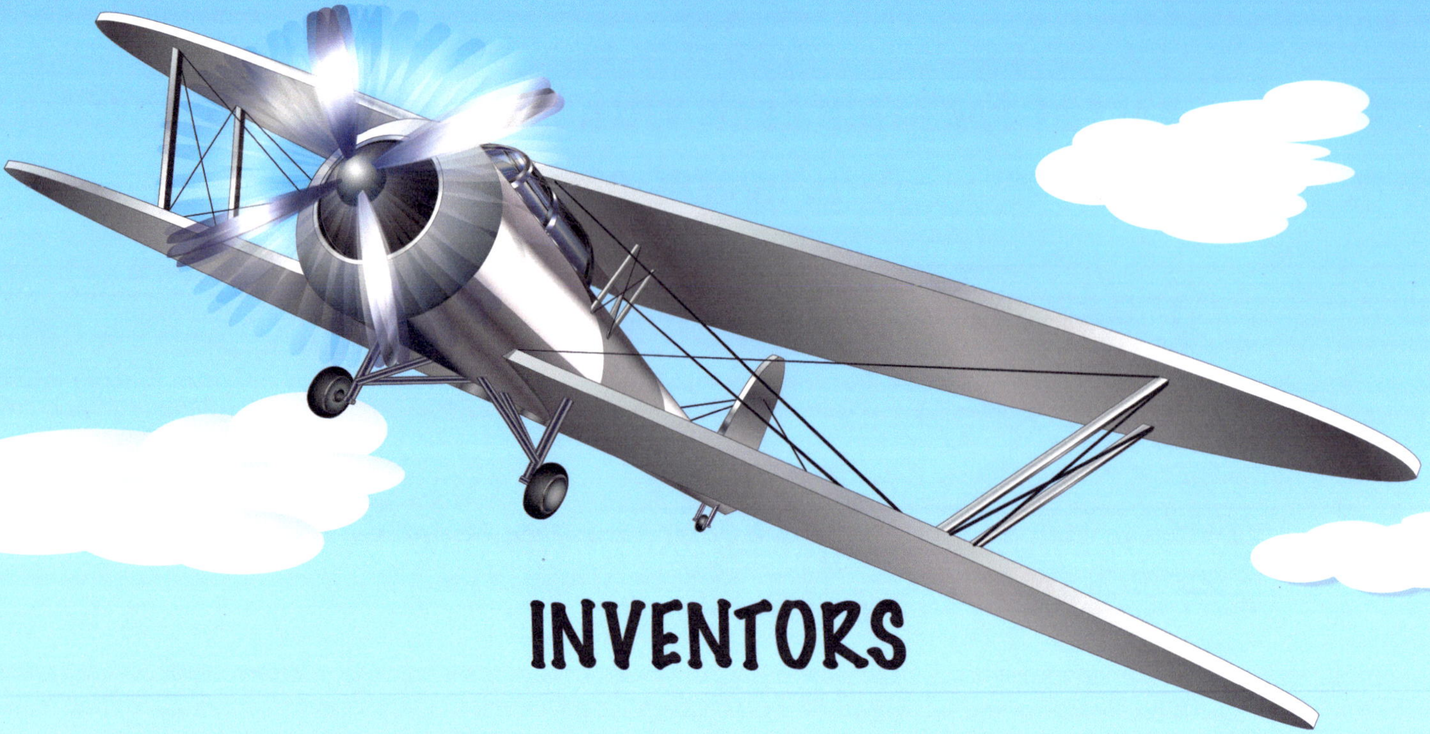

INVENTORS

Q: The Nobel Peace Prize is one the most prestigious award in the field of science. The man behind the award - Alfred Noble invented what?
A: Dynamite

Q: Who invented the airplane?
A: The Wright Brothers

Q: Who invented the first car?
A: Karl Benz

Q: Who invented the first computer?
A: Charles Babbage

Q: Who invented the Credit Card Magnetic Strip?
A: Ron Klein

Q: Who invented the Electric Battery?
A: Alessandro Volta (1800)

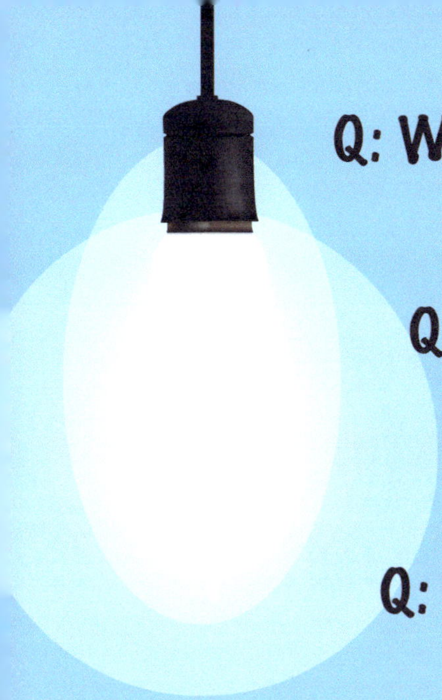

Q: Who invented the Electric Light Bulb?
A: Thomas Alva Edison

Q: Who discovered a way to tread
infections with Penicillin?
A: Alexander Fleming

Q: Who invented the Printing Press?
A: Johannes Gutenber

Q: Who invented scissors?
A: Leonardo da Vinci

Q: Who invented the Steam Engine?
A: Thomas Newcomen (1712)

Q: Who invented the telephone?
A: Alexander Graham Bell

Q: Who invented the first television?
A: Philo Farnsworth

Q: Who invented the Internet (World Wide Web)?
A: Tim Berners-Lee

Q: Botany is the scientific study of what?
A: Plants

Q: A scientist who studies rocks is called a what?
A: A geologist

HOLIDAYS

Q: What is the highest grossing holiday movie of all time?
A: Home Alone

Q: How many nights is Hanukkah celebrated?
A: Eight

Q: What is the Jewish New Year called?
A: Rosh Hashanah

Q: What is the name of the marshmallow Easter treats that come in the shape of birds and bunnies?
A: Peeps

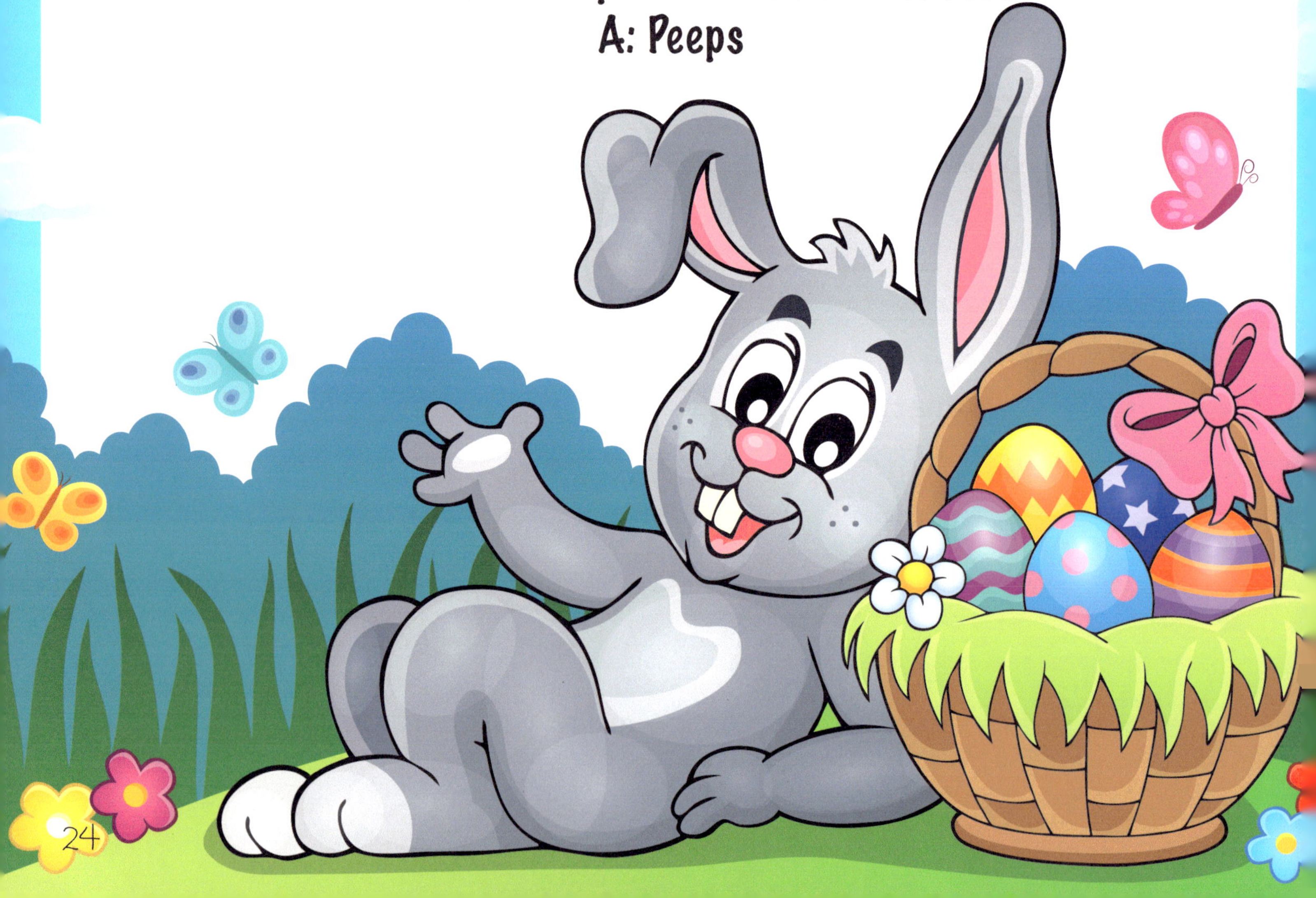

NUMBERS

Q: How many colors are in a rainbow?
A: 7

Q: What is the speed of light (miles per second)?
A: 186,000 miles per second

Q: What is the speed of sound in mph? Bonus: ft./sec.
A: 761 mph Bonus: (1100 ft./ sec.)

Q: What is the total if you add up all the numbers from
1 to 100 consecutively (1+2+3...)
A: 5050

Q: How many zeros in a trillion?
A: Twelve zeros

Q: What is the next named number after a trillion?
A: Quadrillion

Q: What does 111,111,111 x 111,111,111 =?
A: 12,345,678,987,654,321

Q: Fibonacci number sequence strongly relates to
The Golden Ratio such that each number is the sum of two
preceding ones. Starting with 0, 1, 1, 2, 3, 5, 8... next number is?
A: 13

Q: How many dimples on a golf ball?
A: 336 dimples

PLANETS & SPACE

Q: What is the most common element in the Universe?
Bonus: Chemical Symbol.
A: Hydrogen. Bonus: H

Q: Who was the first man to step on the moon?
A: Neil Armstrong

Q: How many planets are in our solar system?
A: Eight

Q: About how many stars are in the Milky Way?
A: 150-250 billion

Q: What color is Mars? Bonus: Why?
A: Red. Bonus: Iron Oxide

Q: Which planet is closest to Earth?
A: Venus (on its orbit). Bonus Mercury is closest to Earth the longest.

Q: What is the largest planet in our solar system?
A: Jupiter

Q: What planet spins the fastest, completing a rotation in just 10 hours?
A: Jupiter

Q: What planet is the hottest in the solar system?
A: Venus

Q: Which planet in our Solar System is known for having a ring?
A: Saturn

Q: What two planets literally rain diamonds?
A: Jupiter and Saturn

Q: Can you name the closest star to Earth?
A: The sun

Q: How long does it take for light to travel from the Sun's surface to the Earth?
A: 8 minutes and 20 seconds

Q: Which planet has a day that lasts almost eight months on Earth?
A: Venus

Q: What was the first animal to go into orbit? Bonus: Name?
A: A dog. Bonus: What was the name of the dog? Laika

Q: How many Earths can fit inside the sun?
A: 1.3 million

PLANTS

Q: How many plants species are there?
A: Over 300,000

Q: How many species of edible plants are there on Earth?
A: More than 80,000

Q: 90% of the foods humans eat come from how many plants?
A: 30

Q: Medicine utilizes how many different species of plants?
A: 70,000

Q: What percent of plant life is found in the ocean?
A: 85%

Q: The Amazon Rainforest produces what percent of the world's oxygen supply?
A: 50%

Q: What is the name of the tallest tree ever?
Bonus: How tall?
A: Australian Eucalyptus. Bonus: 435 feet tall!

Q: The average size tree can make how many pencils? Bonus: Average height is what?
A: 170,100 pencils. Bonus: 87.6 feet tall.
Bonus 2: 435 feet = 844,674 pencils

Q: Is Cucumber a fruit or vegetable? Bonus: Why?
A: Fruit. Bonus: Because it has seeds in the middle

Q: What is the tallest type of grass?
Bonus: How many inches can it grow in a day?
A: Bamboo. Bonus: 35 inches in a day!

Q: Where are all the vitamins in a potato?
A: The skin.

Q: What is the most violently toxic plant in North America?
A: Water Hemlock (Cicuta Masculata)

Q: What plant was responsible for the death of Abraham Lincoln's mother, Nancy Hanks?
A: White Snakeroot (Ageratina Altissima).
Nancy Hanks was poisoned by simply drinking the milk of a cow that had grazed on the plant.

Q: What plant kills the most people per year?
Bonus: How many die a year?
A: Tobacco (Nicotina tabacum). Bonus: 5 million deaths a year. The leaves contain a toxic alkaloids nicotine and anabasine that can be fatal if eaten. Nicotine is both psychoactive and addictive.

POOP

Q: What is bat guano (poop) explosive? Bonus: How was it used?
A: Consists of Potassium Nitrate (Saltpeter). Bonus: It has been harvested for making gunpowder.

Q: What product does the Civet produces?
A: Most expensive coffee beans.

Q: Which animal is in a constant state of pooping and vomiting? Bonus: Why?
A: House Fly. Bonus: When it eats, it immediately digests and goes right back out. Since the frontal excrements are a way of pre-digesting food before they consume it. If a fly lands on you, it has already pooped and vomited on you.

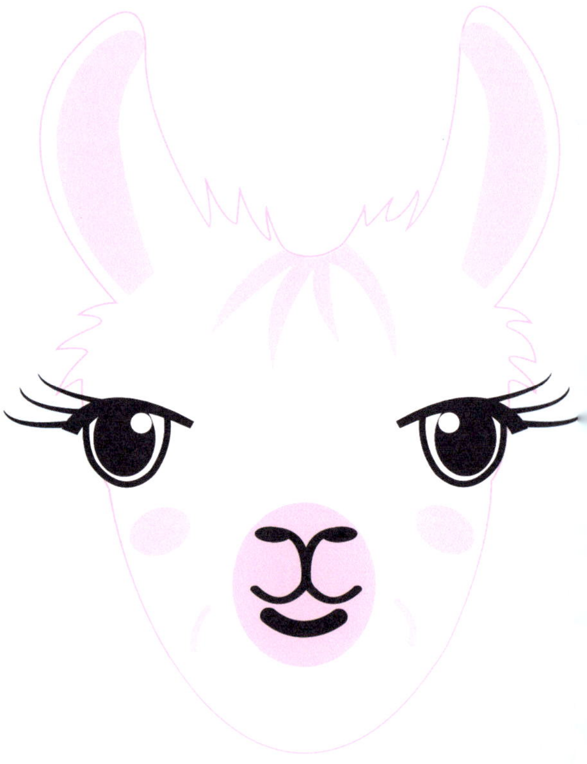

Q: What animal has a special type of poop that fights pollution? Bonus: Why?
A: Llamas. Bonus: It dilutes acids that comes from mines and also dissolves metals that could be harmful when consumed.

Q: How much paper is produced a day from elephant dung?
A: Elephant dung can produce 100 pages of pager a day (POO-APER)

Q: Who left poop on the moon? Bonus: How much poop?
A: Neil Armstrong Bonus: He was kind enough to leave four bags of poop.

Q: What animal poops sand? Bonus: Why?
A: Parrot Fish. Bonus: Because they eat coral with their beaks.

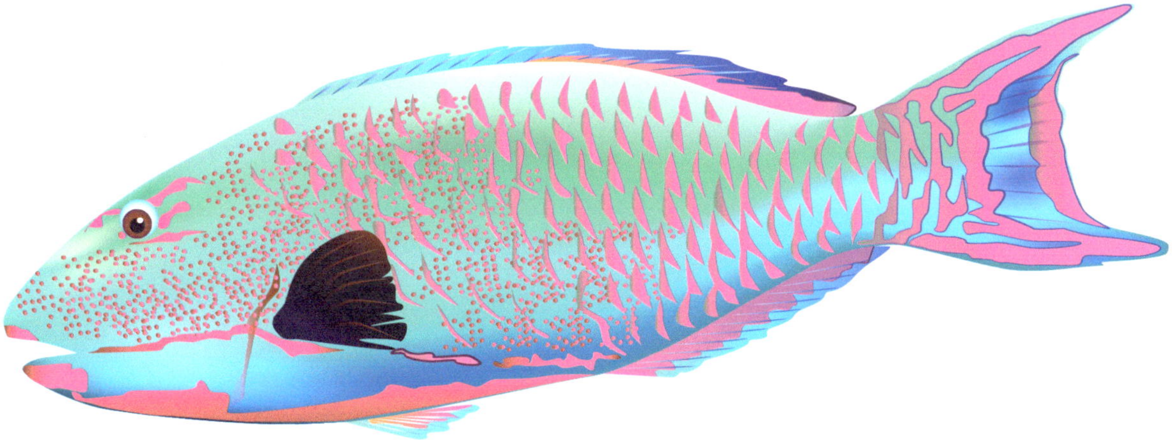

PSYCHOLOGY

Q: What is the phenomenon that explains why people will tend to refuse to offer help when other people are present during an emergency?
A: The Bystander Effect

SCIENCE & TECHNOLOGY

Q: What temperature does water freeze?
A: 32° F

Q: What temperature does water boil?
A: 212° F

Q: What is the molecular structure of water?
A: H20

Q: Water has a pH level of around
A: 7

Q: How many elements are in the periodic table?
A: 118

Q: Diamonds and Graphite are both
made entirely of what element?
A: Carbon, C

Q: Most of a penny is made up from what type of metal?
A: Zinc

Q: Who was the scientist to propose the three laws of motion
A: Sir Isaac Newton

Q: Who invented the modern periodic table?
A: Dmitri Mendeleyev

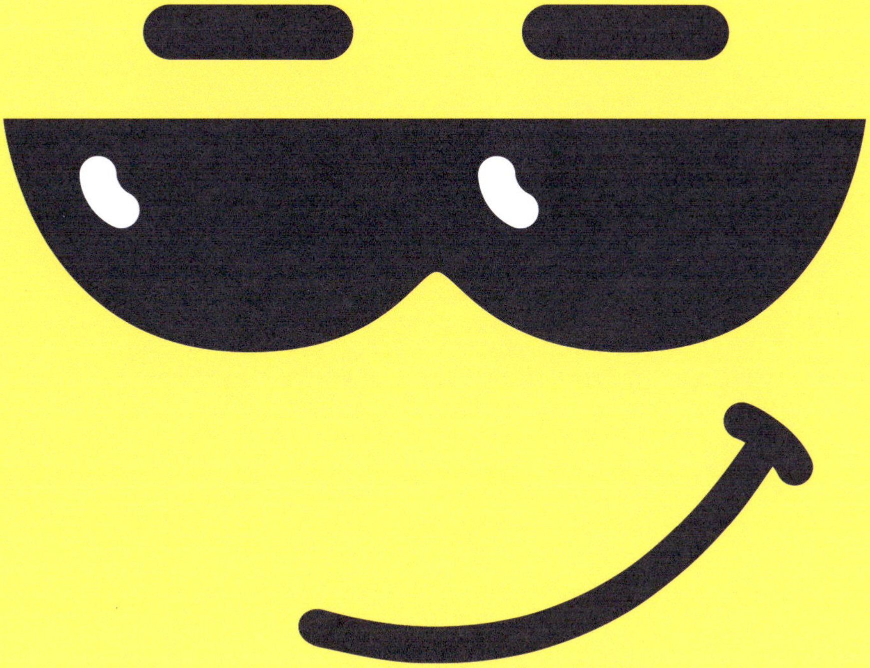

SMILES

Q: What chemicals are released from your brain when you smile?
A: Dopamine, oxytocin, serotonin and endorphins

Q: When is National Keep Smiling Day?
A: May 31st

Q: Who smiles more, a men or a women?
Bonus: How often for each?
A: Women. Women smile an average of 62 times a day, Men, 8 times.

SPORTS

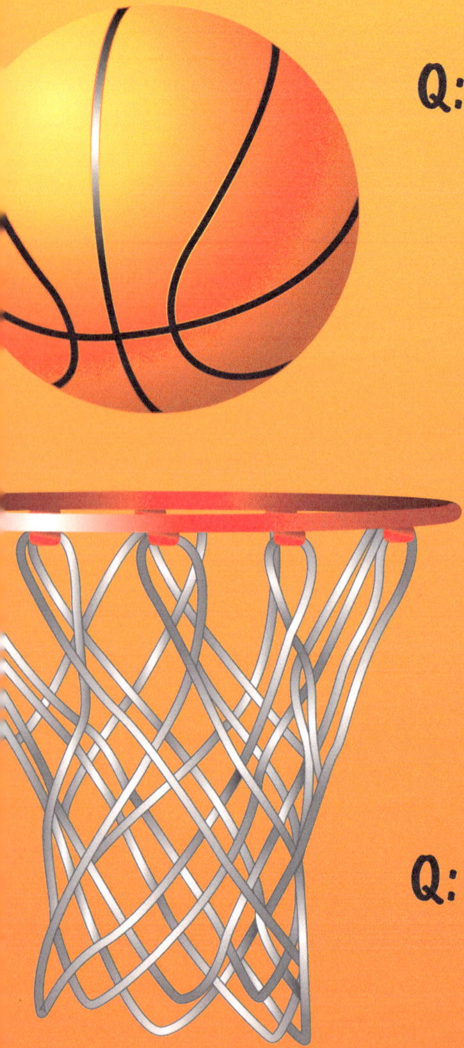

Q: Where country did the Olympic games originate?
A: Athens, Greece

Q: How many rings make up the Olympic rings?
A: Five. Bonus! Name the five colors.
Blue, yellow, black, green, red.

Q: What country won the very first
FIFA World Cup in 1930?
A: Uruguay

Q: In what sport can you get a hole in one?
A: Golf

Q: How big is the diameter of a basketball hoop?
A: 18 inches

Q: In sports, what is an MVP?
A: Most Valuable Player

Q: What famous horse won the
Triple Crown in 1973?
A: Secretariat

WEATHER

Q: Stratus, cirrus, cumulus and nimbus are types of what?
A: Clouds

Q: What does a thermometer measure?
A: Temperature

Q: During a thunderstorm, which comes first: lightning or thunder?
A: Both. They occur at approximately the same time. However, if we're at a distance from a storm, we see lightning before we hear thunder because light travels much faster than sound.

*****

Courtney McLaughlin is a freelance writer and editor in Charlotte, NC who loves a good game of trivia. She excels in the Arts & Entertainment category and is dismal in Sports.

ABBREVIATIONS

Q: What does H20 stand for?
A: Water

Q: What does 'EPCOT' in Disney's
EPCOT Center stand for?
A: Experimental Prototype Community
Of Tomorrow

Q: What does LASER stand for?
A: Light Amplification (by) Stimulated Emission
(of) Radiation

Q: What does M&M's Stand for?
A: Mars & Murrie's, referring to the founders
Forest Mars, Sr. and Bruce Murrie

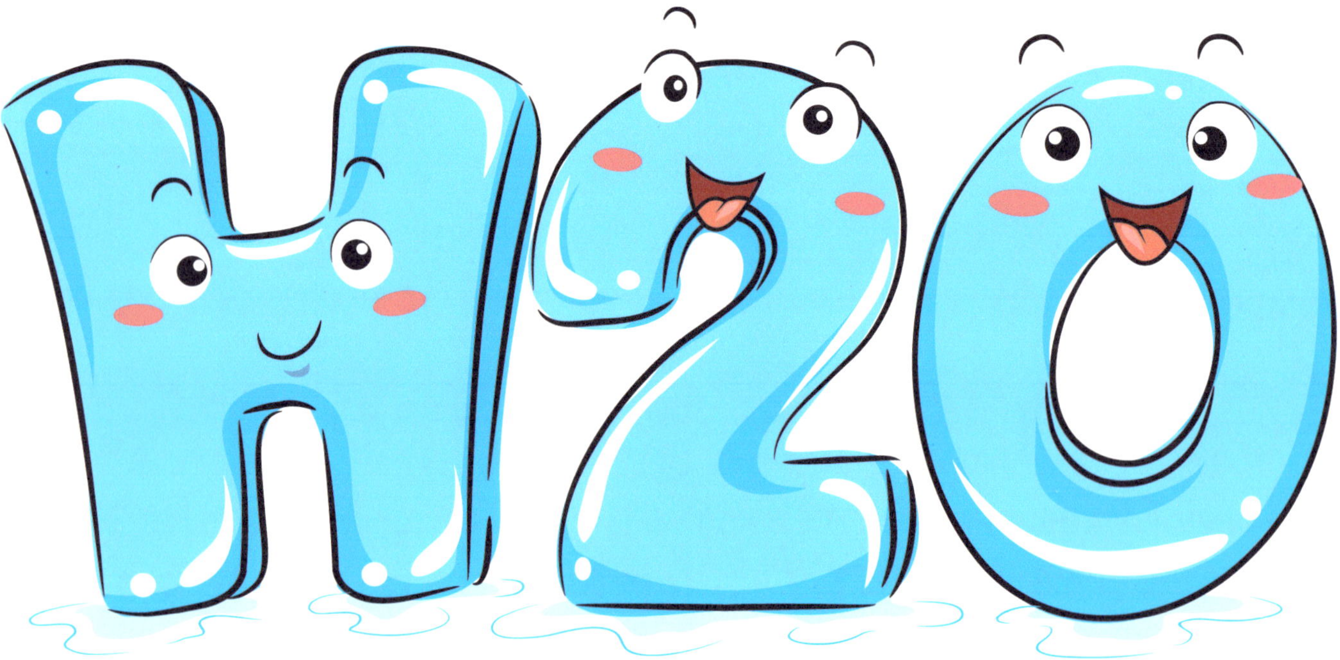

Q: What does NERF stand for?
A: Non Expandable Recreational Foam

Q: What does 'Smart' stand for in Smart car?
A: Swatch Mercedes Art, because of their collaboration on this project.

Q: What does SPAM stand for?
A: Shoulder Pork And Ham

Q: What does Yahoo! stand for?
A: Yet Another Hierarchical Official Oracle

Q: In a website browser address bar, what does "www" stand for?
A: World Wide Web

Q: What does 'D.A.R.E.' an acronym for?
A: Drug Abuse Resistance Education

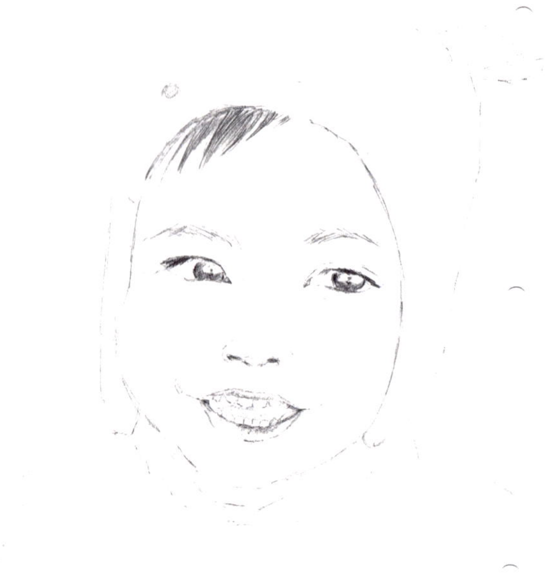

About Kenny

Kenny 'K3' Rochon, III is a 7-year-old boy, Game Changer - Leader. He recently was published as an author, and he is an expert in children's jokes.

When K3 is not telling jokes, he is playing chess with his daddy, learning how to code, playing Minecraft and Nintendo, and even some learning games that challenge his inquisitive and problem-solving mind.

He enjoys math, playing Sudoku, and expanding his vocabulary with his "Kenny's Fortune" word box. Kenny's Future Word Box is a future product he is coming out with the help other children learn the military alphabet and learn positive powerful words. He uses these words to hone up on his skills as a proficient speller so he can win spelling bees.

He has a yellow belt in Tae Kwon Do and practices so he can protect his Mommy (and Daddy) from danger.

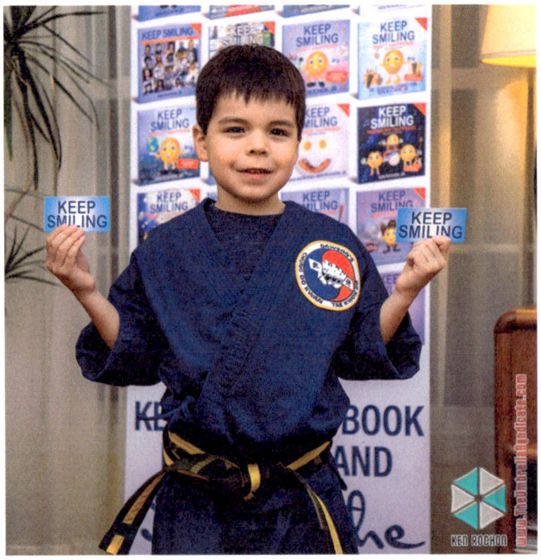

Drawing by Stephania Christianson

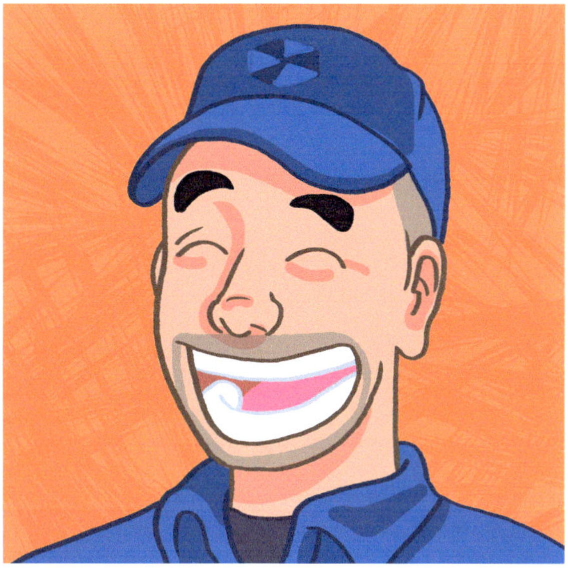

About Ken Rochon

Ken loves his son Kenny. He enjoys writing a new book every year to celebrate his son's life and to share valuable lessons with him on character, leadership, and success

You can listen to Ken on his radio show Amplified with Ken Rochon, through the VoiceAmerica Influencer Channel on Mondays at 11 AM Eastern, on Itunes or at www.AmplifiedRadioShow.com

This is Ken's 28th book and now his favorite :) He wants to inspire other parents to create a legacy of love with their children through publishing. Ken has created templates that make publishing books more effortless and affordable. (www.PerfectPublishing.com)

Besides playing with his son, Ken loves to travel the world and has been to over 100 countries. He is a world class photographer and enjoys covering the biggest events in the USA (www.BIGeventsUSA.com)

If you are interested in having Ken speak at your school or event, simply connect with him on Facebook or email him at Ken@TheUmbrellaSyndicate.com.

Ken is the co-founder of the Keep Smiling Movement.com. He has co-authored a series of books focused on setting your mindset to be positive. Check out on Amazon the *Keep Smiling – Shift Happens* Series.

He resides in the Baltimore Washington Metropolitan Area with his family.

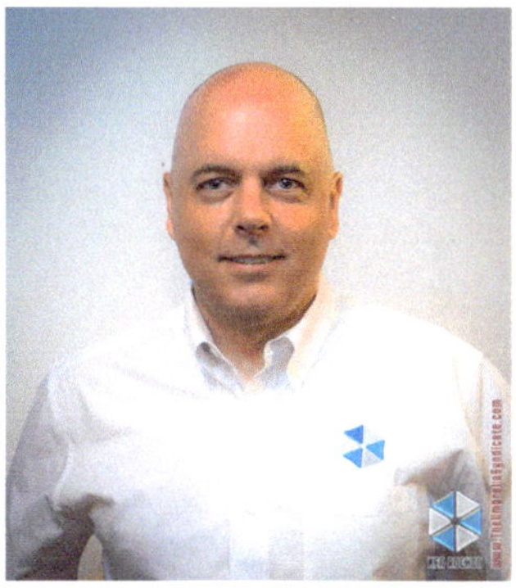

More Books From Kenny

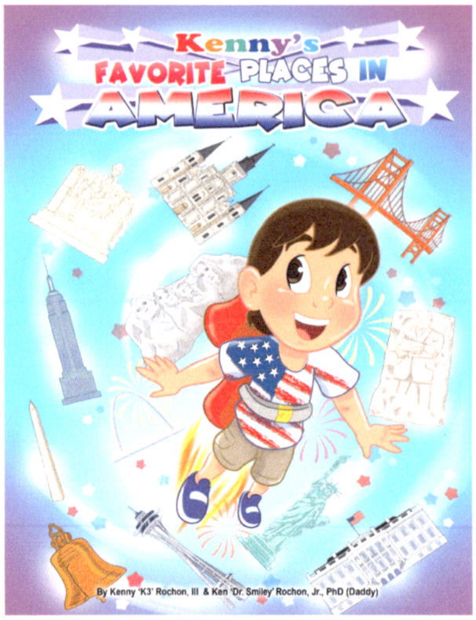

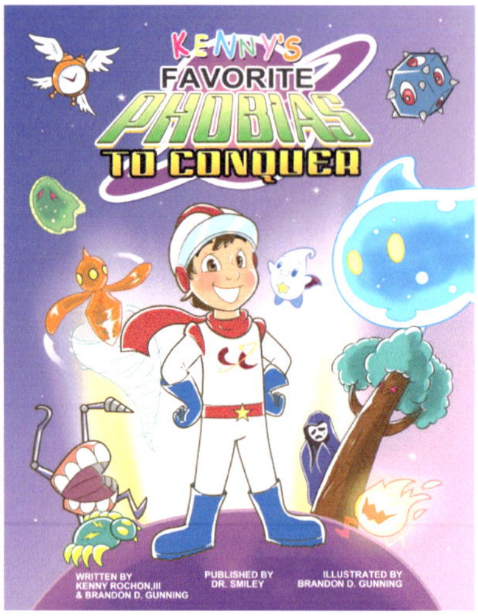

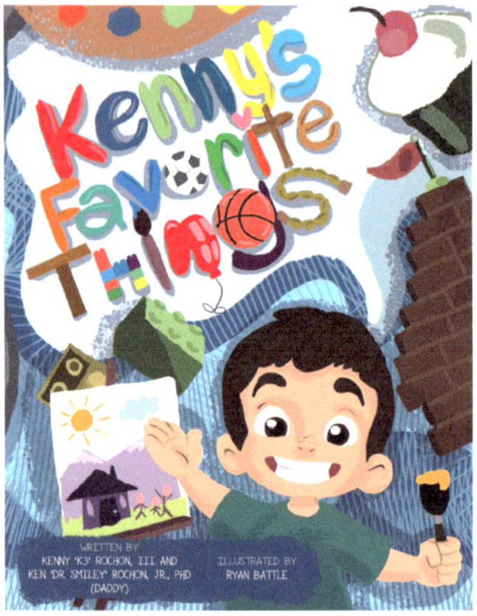

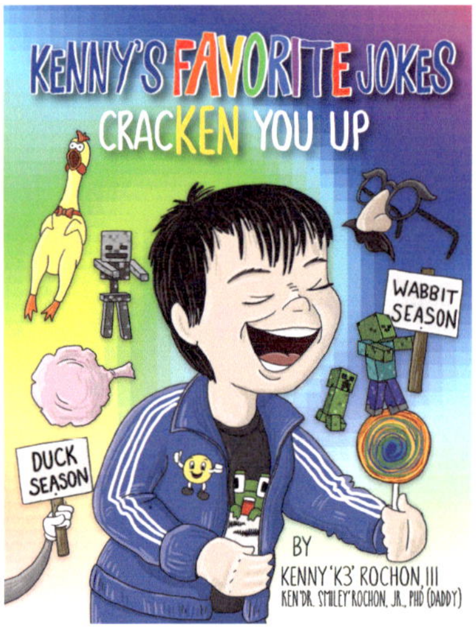

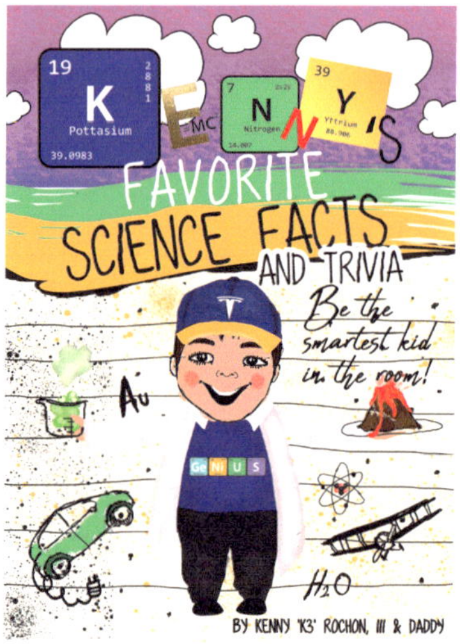

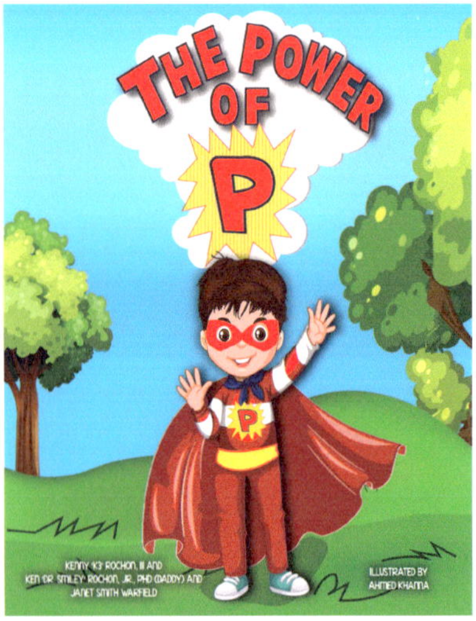

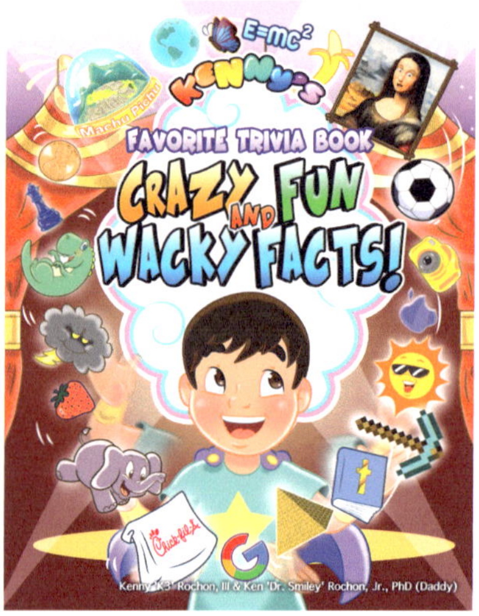

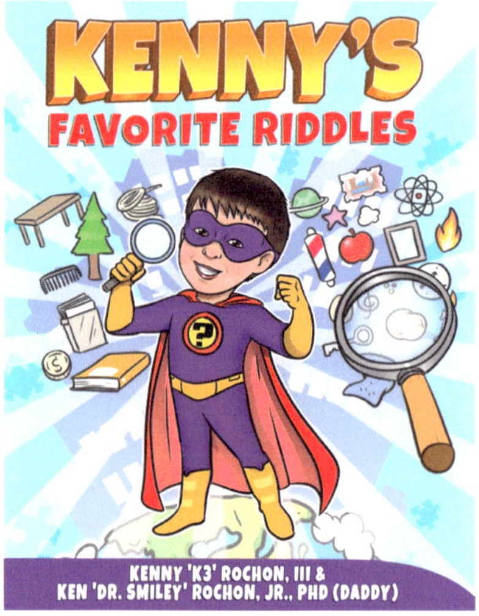

www.PerfectPublishing.com

Made in the USA
Monee, IL
05 November 2026

67383296R00029